big nose
proboscis monkey

long nose
anteater

warm nose
bunny

red nose
mandrill

wet nose
seal

pink nose
kitten

dry nose
chameleon

for Chris and Carol, for all the
years of friendship, love, and strength—ST

For Leanne and Emilia, my inspiration and my world—MG

Library of Congress Cataloging-in-Publication data is on file with the publisher.

Illustrations by Michael Garton
First published in the United States of America in 2019 by Albert Whitman & Company
ISBN 978-0-8075-9046-1 (hardcover)
ISBN 978-0-8075-9047-8 (ebook)

Printed in China
10 9 8 7 6 5 4 3 2 1 HH 24 23 22 21 20 19

Design by Aphee Messer

For more information about Albert Whitman & Company,
visit our website at www.albertwhitman.com.

100 Years of Albert Whitman & Company
Celebrate with us in 2019!

Whose Are These?

Whose Nose?

Sue Tarsky illustrated by Michael Garton

Albert Whitman & Company
Chicago, Illinois

big nose

tiny nose

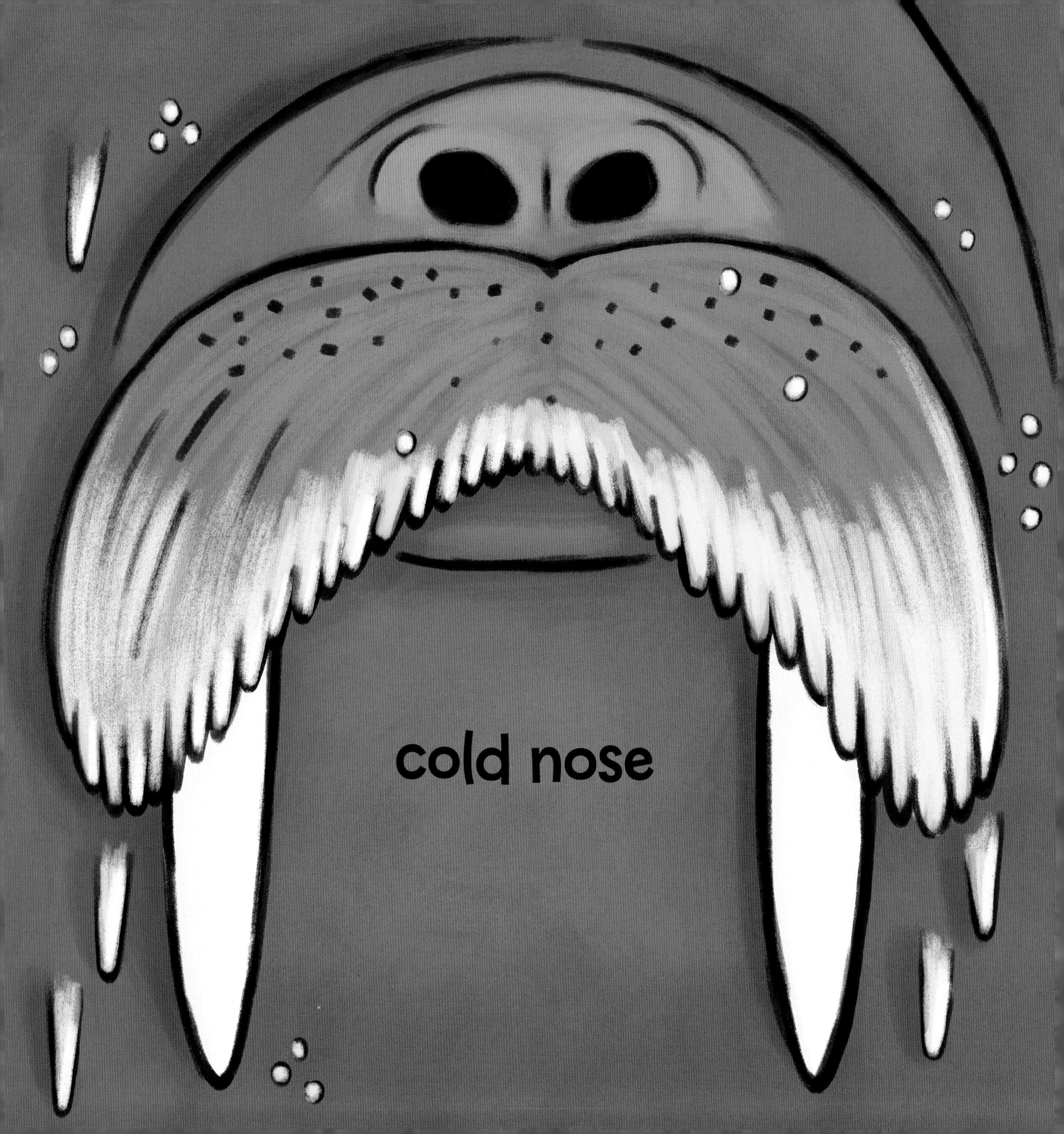
cold nose

warm nose

long nose

short nose

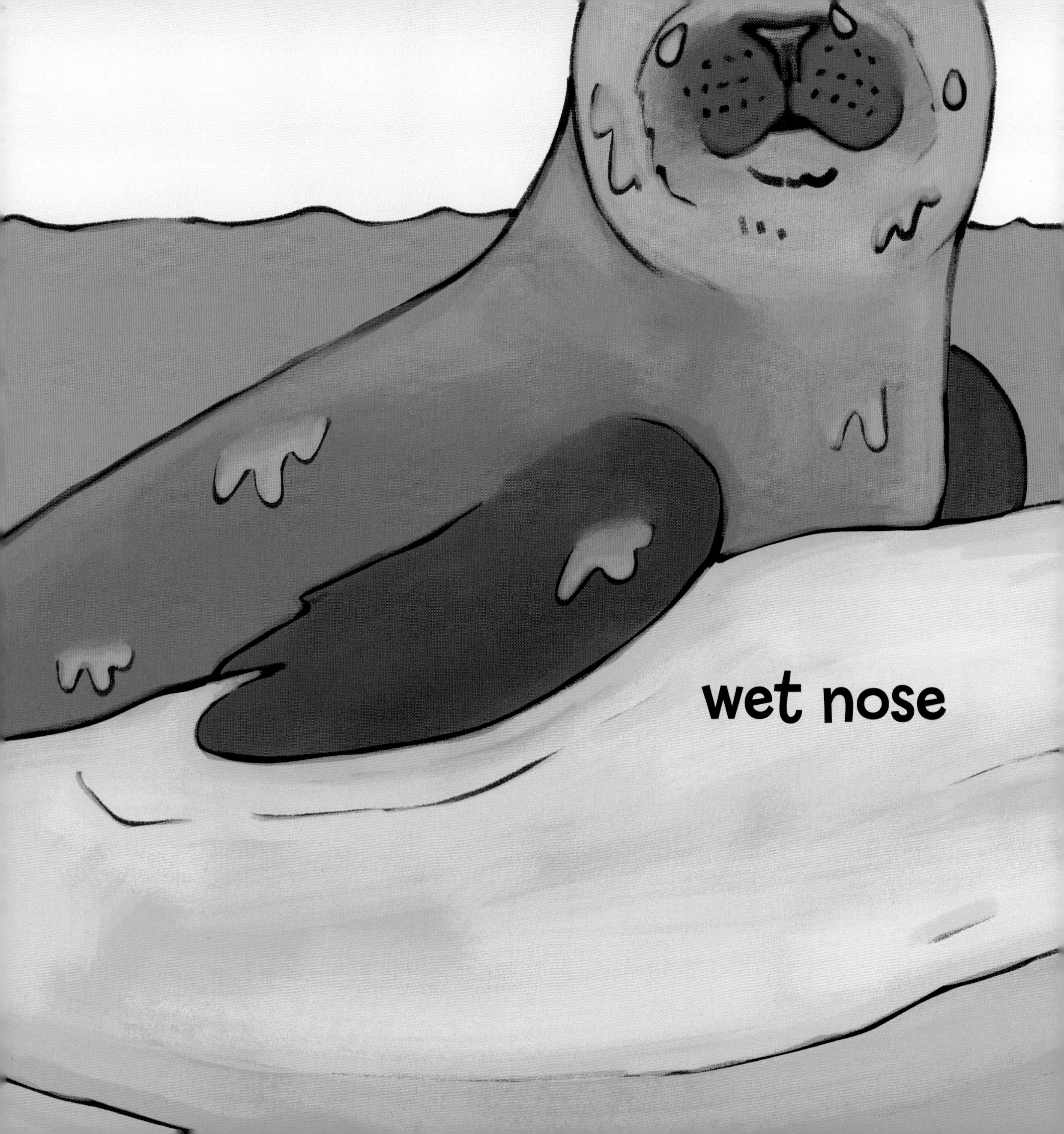
wet nose

dry nose

red nose

pink nose

pointy nose

round nose

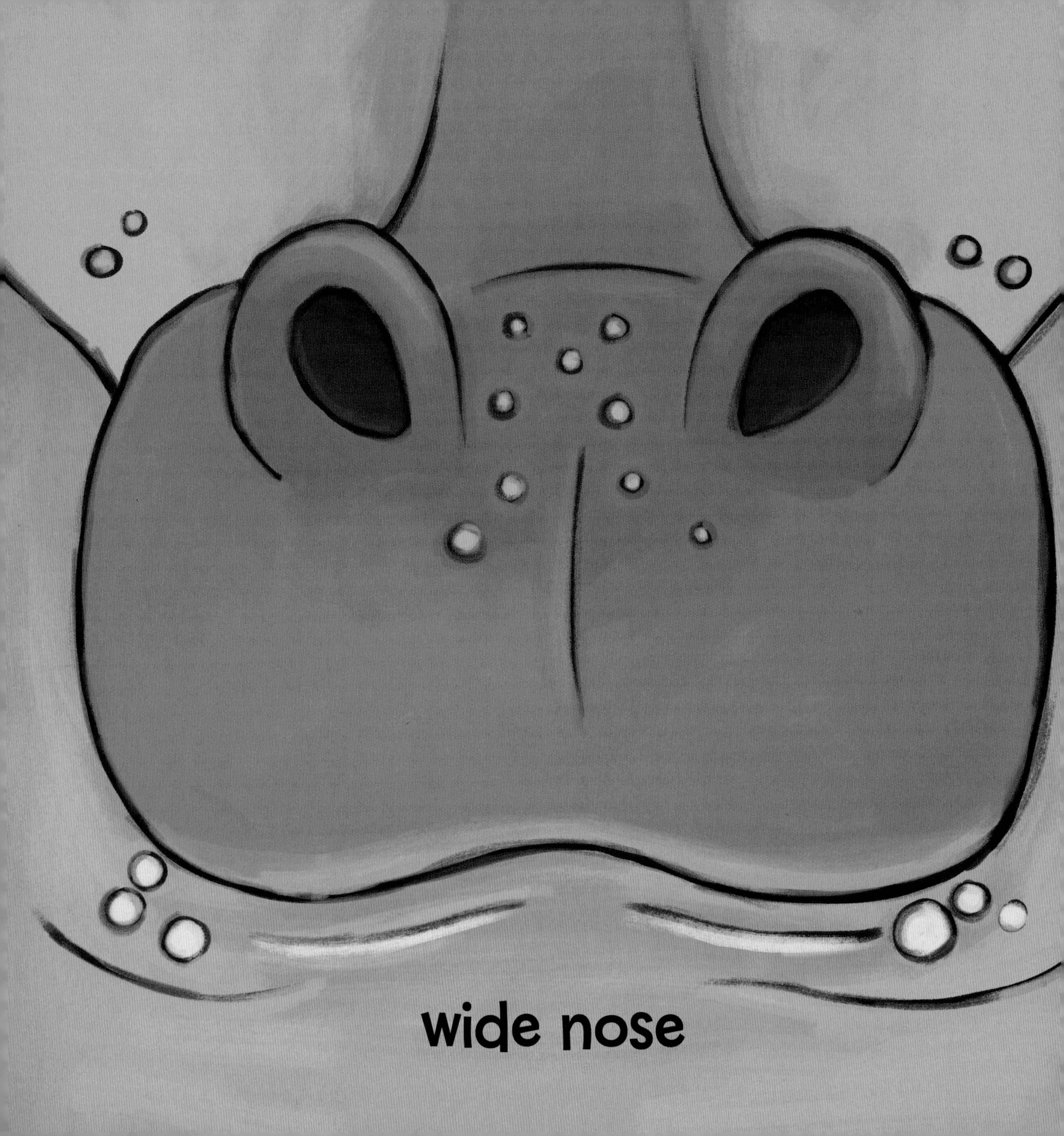
wide nose

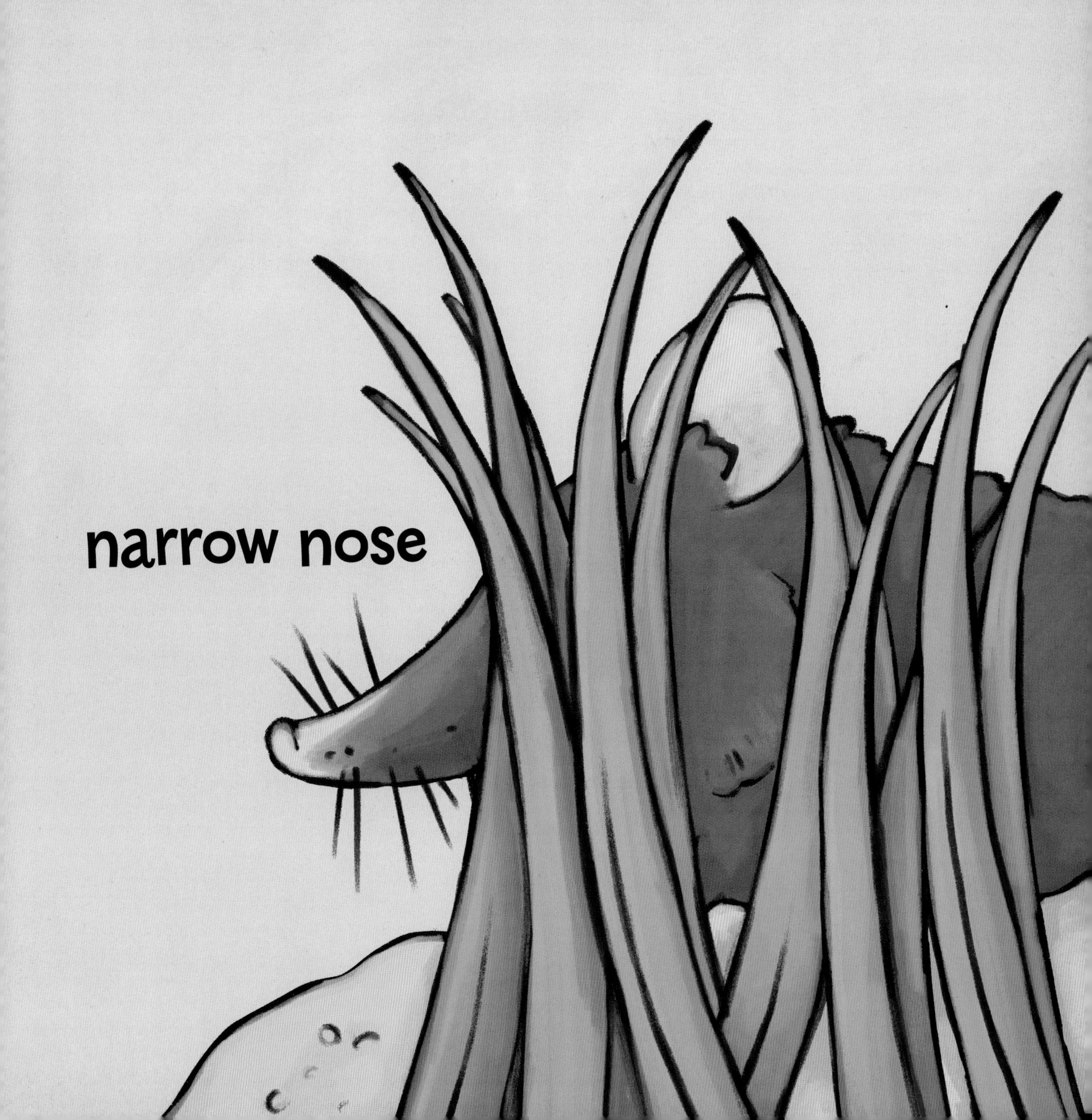
narrow nose

your nose

my nose!

round nose
porcupine

narrow nose
shrew

tiny nose
mouse

cold nose
walrus

wide nose
hippo

short nose
pug dog

pointy nose
crocodile